Our solar system is an exciting, act[...]e. It's our neighborhood, our address in th[...]sterious and wonderful universe. One day - a day not too far away - we'll be exploring the other planets, mining metals on asteroids and even taking sightseeing trips to the moon. But before we do all of that, it's a good idea for us to get to know the neighbors.

So let's take a trip around the solar system and check out our corner of space that's jam-packed with adventure and action: visit a world where it rains diamonds, call in on a moon where you can flap your arms and fly, and zoom past a planet hot enough to cook a pizza in 7 seconds.

And while we're touring the planets, you might find yourself dreaming about being someone who will wrap their hands around the future and help to take us there.

Dedicated to my grandsons Colin and Ryder, two young men
who are already wrapping their hands around the future

MERCURY

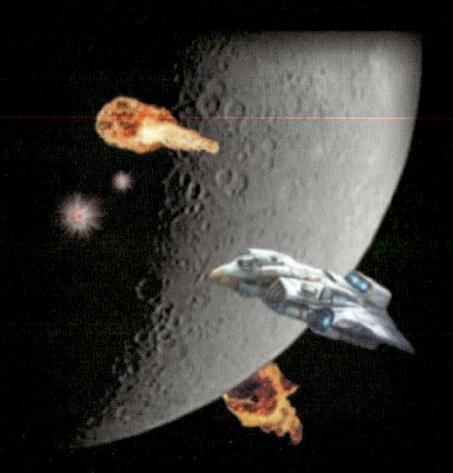

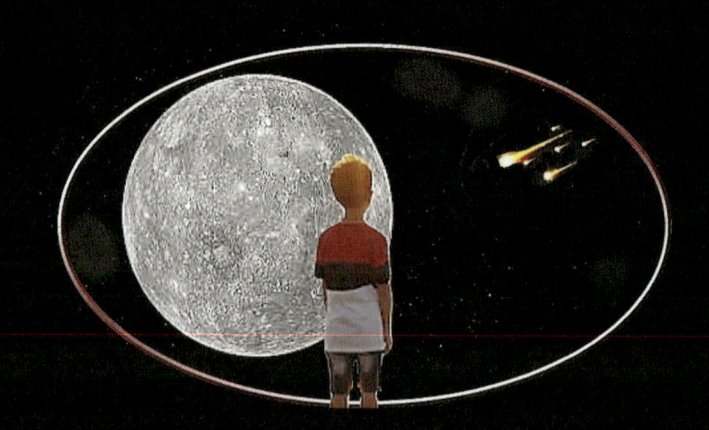

A PRETTY ROUGH AREA

The sun's huge gravity acts like a magnet for space junk and a lot of this smashes into Mercury, making it the most cratered planet in the solar system. The biggest crater of them all is the Caloris Basin. About 4 billion years ago, something crashed into Mercury, making a hole 960 miles (1,550 kms) wide, big enough to fit all of Texas inside.

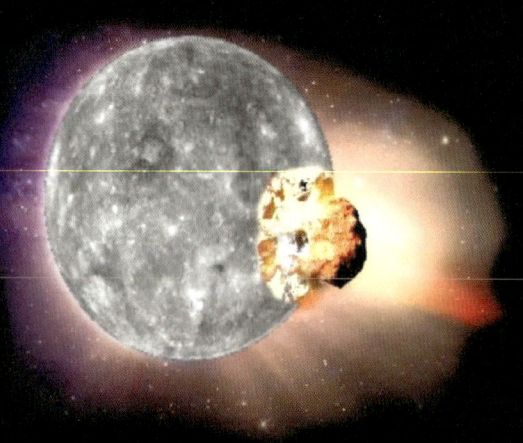

NOT THE HOTTEST PLANET

Even though it's the closest planet to the sun, Mercury *isn't* the hottest. Venus is. Mercury keeps one side facing the sun, a side that gets as hot as 800º F. On the other side of the planet, the temperature is way down at minus 300º F. So there's a hot side and a cold side, with no warm bit in between.

ICE ON THE HOT SIDE?

It's a bit complicated – chemistry and all that – but on the inside of the crater walls in the north and south poles of the side facing the sun, there's ice. Proper ice.

THE MAGNIFICENT SEVEN

America's very first space program was named after the planet Mercury. It began in 1958 (that's when NASA first started) and ended in 1964. Seven astronauts – who should be called the Magnificent Seven – flew into space as the true pioneers of the space age. Their monument at Cape Canaveral is a number 7 inside the symbol for the planet Mercury.

MESSENGER FROM EARTH

NASA'S Messenger spacecraft was the first probe ever to orbit the planet Mercury. It arrived in 2011 and ended its mission in 2015 with a planned crash into the planet's surface. While it was circling Mercury, Messenger took more than 200,000 photos and pretty much mapped the whole surface.

THE PLANET THAT WASN'T

Over a hundred years ago, scientists thought that because Mercury wobbled in its orbit, there must a planet between it and the sun. They called this planet Vulcan after the Roman god of fire and volcanoes, but it was never found. Many years later, Einstein proved that Mercury wobbled simply because of its very strange orbit. The planet Vulcan had never existed.

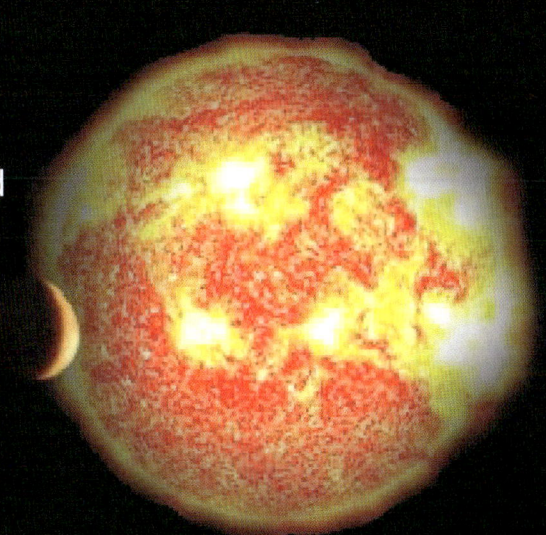

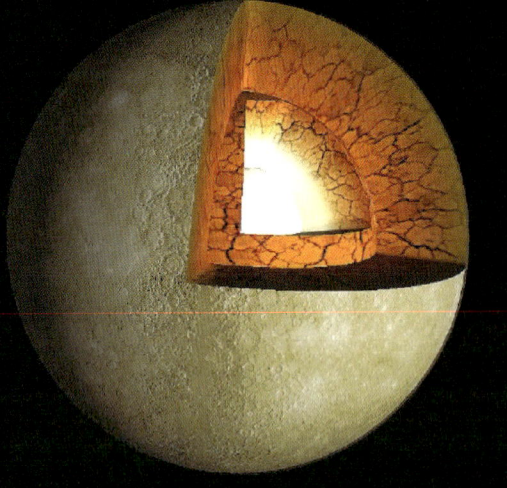

IRON HEART, WRINKLY FACE

Mercury is the most 'wrinkly' of the planets. That's because deep inside, there's a huge core of molten iron which is shifting around a lot.

That means that the surface of Mercury stretches and cracks and forms what are called wrinkle ridges. So, with all those splits and bumps, Mercury is a seriously rough planet to visit.

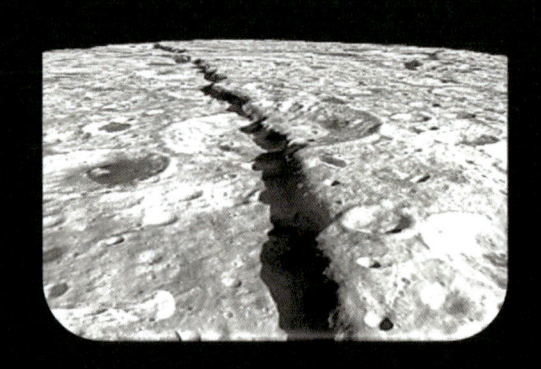

THE GHOSTS ON MERCURY

Can you see the circle in this photo? That's a ghost crater. Millions of years ago a large rock crashed into Mercury and made a huge hole. But over time it was filled with volcanic dust and stuff, leaving only the outside walls showing. A ghost crater.

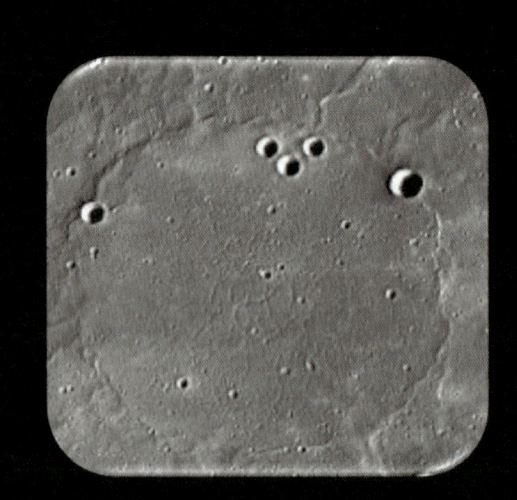

WHY DOES IT LOOK SO GRAY?

Mercury doesn't shine and shimmer like most of the other planets in our solar system. That's because it's covered with a dark, rocky material called graphite. Which is pretty much the same stuff that's in the middle of your pencil.

Distance from the sun - 36 million miles (58 million kms)

Length of Day - 58 Earth days

Length of year - 88 Earth days

Diameter - 3,030 miles (4,876 kms)

Number of moons - None.

THE FLYING GOD

In Roman times Mercury was kind of like a postman, taking messages from one god to another by flying around on his winged feet. Oh, and he was also the god of trickery and thieves.

HOW HIGH COULD I JUMP ?

On Mercury you would only weigh about 40% of what you do here on Earth, so you would be able to jump twice as high. Of course you couldn't breathe because there's no atmosphere, and it would be 800° F on the sunny side, so I'm guessing that jumping around wouldn't be a heap of fun.

WHO DISCOVERED IT ?

Nobody knows who discovered Mercury, but It might have been the Sumerians over 3,000 years ago. Mercury is one of the five planets in our solar system that can be seen without a telescope. But because it's so close to the sun, it's still very hard to spot.

VENUS

OUR SISTER PLANET - NOT

Because Venus is about the same size as Earth and is made of pretty much the same stuff, it's often called our sister planet. But, apart from being too hot to even land on, it's seriously different in other ways

SPACE CRASH!

How did Venus get to be so hot and nasty? Some scientist think that a small planet may have crashed into it millions of years ago and made the fiery wildness that we see today. Such a crash isn't unusual. In fact, it happens quite a lot in our solar system.

THE HOTTEST PLANET

Venus is the hottest planet in the solar system, with over 37 active volcanoes spewing molten lava all over its surface. Add to that the lightning that never, ever stops, and you've got a recipe for a planet that's a real nightmare to visit.

A SEVEN SECOND PIZZA?

Venus is so hot (**900°F**) that it would only take 7 seconds to cook a pizza. Part of the reason is that the atmosphere is so dense, heat can't escape into space.

BAD AIR, REAL BAD AIR

And if that doesn't put you off visiting, what about the air that's full of sulfuric acid and is so thick that you can't see through it? And the lightning all day and night? This is what the sun might look like from Venus.

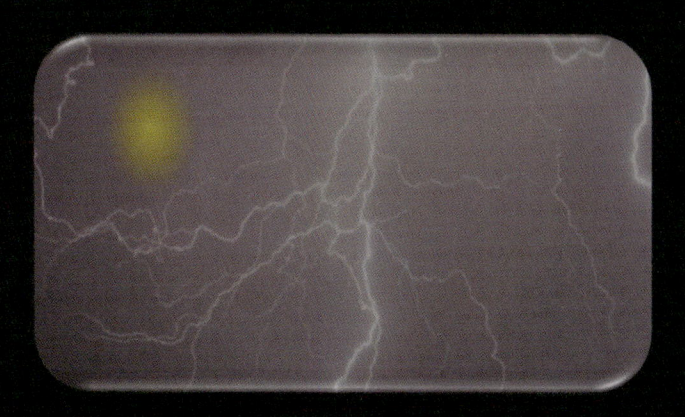

UP, UP AND AWAY

If it's so bad, will we ever be able to explore the surface? The clever men and women at NASA have thought about that. They think that we could stay high up in the atmosphere by using really high-tech balloons to keep safe above all the nastiness below.

YOU CAN VISIT, BUT CAN'T STAY

We've sent quite a few spacecraft to Venus, but mostly just to fly past and take photos. The last one was Venus Messenger. It finished its mission in 2013 after taking thousands of photographs and mapping the surface.

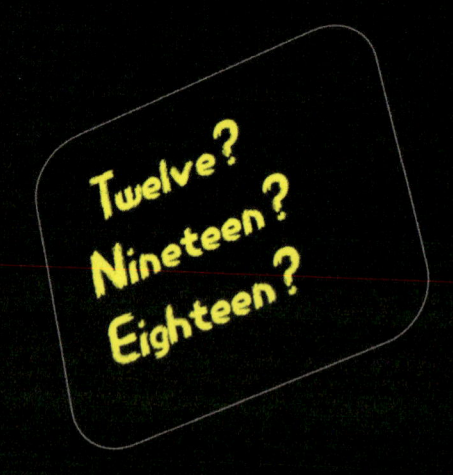

HOW OLD WOULD I BE IF I WAS 12?

A day on Venus is longer than a year. So if you were 12 on Earth, you would be 19 years old on Venus. But weirdly, only 18 days old at the same time. Which is totally confusing and would make birthday parties really hard to organise. The day is so long because Venus is the slowest spinning planet in the solar system.

Twelve? Nineteen? Eighteen?

WHAT WOULD I WEIGH ON VENUS?

Venus' atmosphere - as well as being full of sulfuric acid and all sorts of other nasties - is 90 times heavier than Earth's. So take your weight, multiply it by 90, and there's your answer.

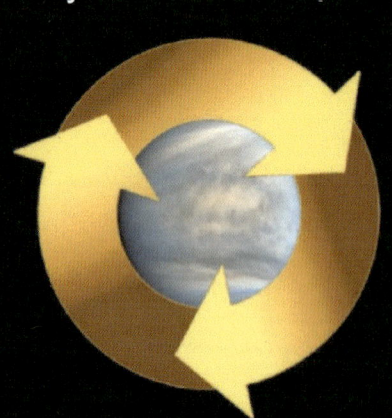

x 90

THE BACKWARDS PLANET

Venus spins in the opposite direction to all the other planets. Which means that when you got up in the morning, the sun would be on the other side of the sky in the West, not in the East. Some scientists think that it spins back to front because the planet is upside down!

CRAZY WINDY

Winds roar across Venus at super-fast speeds that can reach 450 miles (730 kms) an hour. That makes Venusian winds faster than the fastest tornados on Earth.

Distance from the sun - 67 million miles (108 million kms)

Length of Day - 243 Earth days

Length of year - 224 Earth days

Diameter - 7,520 miles (12,100 kms)

Number of moons - 0

THE GODDESS OF LOVE

Venus is the only planet to be named after a woman, which hardly seems fair, does it? The Greeks thought the planet looked so beautiful in the sky that they called it Aphrodite after their goddess of love. The Romans changed that name to Venus – their own goddess of love.

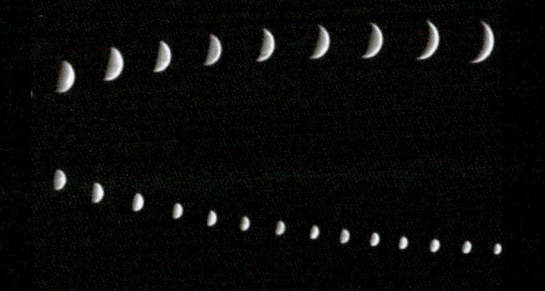

JUST LIKE OUR MOON

Venus has phases just like our moon. You can see them even with quite a small telescope. That's why this jewel in our night sky is a great way to start learning to star gaze.

MORNING AND EVENING STAR

Our sister planet graces both our early mornings and evenings with her glory. The ancient Greeks thought Venus was actually two planets. When they saw her in the morning they called her, *'the bringer of light'* and when they saw her in the evening they called her, *'the star of the evening'*.

MARS

CANALS ON MARS?

In the early 20th century, people believed that there were canals on Mars. They even made maps of these 'canals' and gave them names. This is one of them, which isn't too bad when you compare it with the photo above.

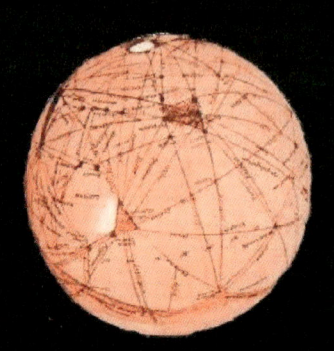

HE STARTED IT ALL

In the late 19th century, Percival Lowell claimed that he could see canals on Mars. Professional astronomers (he wasn't one) didn't take him seriously and laughed at his ideas. But a lot of people didn't, and the myth of scary Martians was born.

THE MARTIANS ARE COMING !

If there were canals on Mars, then someone (or something) must have built them, right? But what if our neighbors weren't friendly? This magazine cover is about a famous book called 'War of the Worlds'. The Martians were very bad neighbors in this, but exciting to read about.

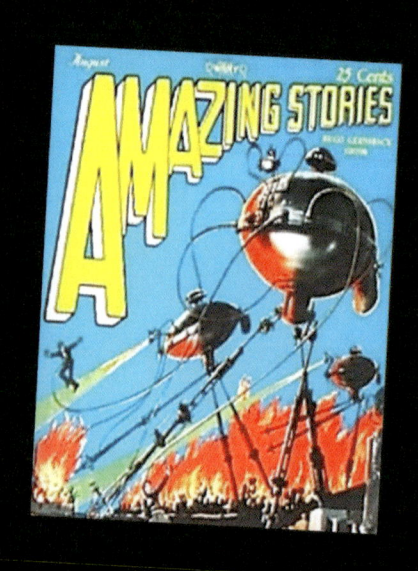

IT'S THE HIGHEST

The highest mountain in the whole solar system is on Mars. It's called Mons Olympus (Mountain of the Gods) and it's 16 miles (26 kms) high. That's right, 16 miles - 3 times higher than Mt Everest. It's so big that all of Hawaii would fit on the its top.

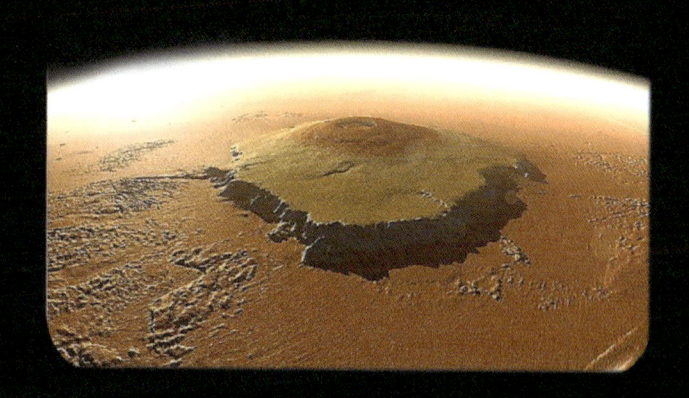

MORE THAN ON THE MOON

Because Mars has such a thin atmosphere (only 1% of Earth's),meteorites don't burn up before they smash into the ground. That's why Mars has a such huge amount of craters, even more than on our Moon. In fact there are more than 43,000 of them – and that's just the ones that are bigger than 3 (5 kms) miles across.

DANGER ! DANGER!

Dust storms are a big deal on Mars. The air is so thin and the red dust is so light, that the wind can easily pick it up and make huge storms that can cover the whole planet. When we go to live on Mars, these storms are going to be something we have to be ready for and learn to deal with.

ISN'T THIS BEAUTIFUL?

Mars is the only planet other than Earth that has a north pole and a south pole. Which is good, because ice means water and water means that we can set up colonies a lot easier. This is the Korolev crater, filled with ice and looking glorious. It's over 50 miles (80 kms) across and deeper than the Grand Canyon.

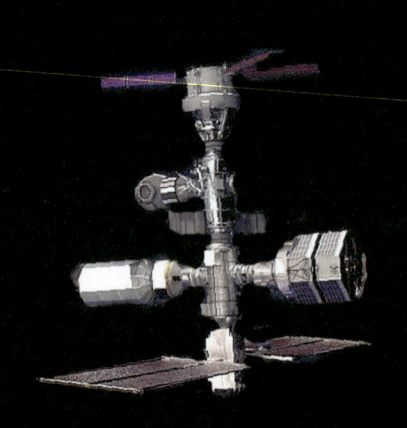

TO LIVE ON MARS, WE NEED TO GO TO THE MOON FIRST

NASA is already planning a mission for people to live on the moon. This a picture of the Gateway spaceship which will take supplies to the astronauts and deliver them down to the Moon's surface. Once we've got the technology for living on the moon figured out, then we're on our way to Mars!

THE FIRST COLONY

Mars will be the first *planet* that we set up a colony on. Sure, we'll live on the Moon first, but that's kind of like a part of Earth, it isn't another planet. So our first full time visit to one of our neighbors in space will be to Mars. Look forward to it.

*Distance from the sun – **142 million miles (228 million kms)***

*Length of Day - **24 ½ hours***

*Length of year - **687 Earth days***

*Diameter - **4,222 miles (6,800 kms)***

*Number of moons - **2***

THE GOD OF WAR

Because of its blood red color, Mars has been called the god of war. The Greeks and Romans thought of the planet as being only second to Jupiter, the king of the gods. Funnily enough, Mars was also the god of farming.

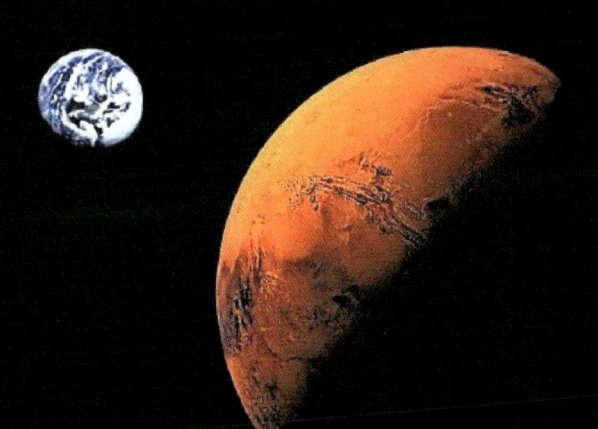

BLUE EARTH, RED MARS

Earth hangs blue and beautiful in space because of all of its oceans and Mars shines red because of all the iron in its rocks. I guess another name for it could be the rusty planet.

TWO SERIOUSLY WEIRD MOONS

Phobos and Deimos are the two moons of Mars and they're a bit odd. Phobos is only 14 miles (23 kms) across and Deimos is just 8 miles (13 kms). Phobos whizzes around Mars three times a day and Deimos does it once. It's like they're playing tag. They've got spooky

ASTEROIDS

A BELT OF ASTEROIDS

Between the orbits of Mars and Jupiter lies the asteroid belt. It's made up of rocks that are as big as 600 miles (970 kms) long and as small as only 20 feet across. But it's not crowded, not like in the movies. The average distance between asteroids is three times that of the Earth to the moon.

DEAD COMETS

Some asteroids are the remains of comets that have had all their ice blown away by the heat of the sun. When it's gone, what's left is a big hunk of rock - an asteroid. But they aren't all made in the same way, and that's what makes the asteroid belt so interesting. How were over 800,000 asteroids made? Maybe an exploding planet? No one knows for sure.

HOW MANY ARE THERE ?

Way too many to count is the answer. At the moment we think that there are at least 800,000, but thousands more are being discovered every year. Of all the places in the solar system, the asteroid belt is the busiest for us to explore.

BOMBARDMENT !

Do we need to worry about an asteroid smashing into our Earth? Yes. The risk is very small, but it's there. In fact, NASA has created the Planetary Defense Coordination Office, which has the job of looking out for asteroids or comets that might collide with us and coming up with a plan of action If they find one.

If we do spot asteroids heading our way, one of the ways to deal with them would be send up missiles to destroy them, or at least to knock them off course.

About 66 million years ago, an asteroid *did* smash into the Earth. It was only 9 miles (14.5 kms) wide, but it still destroyed ¾ of all the animals and plants on our planet. Scientists think that's the reason why the dinosaurs became extinct.

WHOA! THAT WAS CLOSE!

An asteroid whizzed past our Earth on Friday 13th November 2020. It was about the size of a small house. Which doesn't sound very big, but if it had of hit us, it would have caused a world of damage.

THE BIGGEST OF THEM ALL

Deep in the asteroid belt is Ceres. It's 600 miles (1500) across, so it's what's called a dwarf planet. Ceres is the 25th biggest body in the solar system. That big white spot? Not ice, but probably a sort of salt in the middle of the crater - maybe left behind when water boiled away into space.

COPS IN SPACE?

In the 18th century, there was a bunch of people who figured out that there was a missing planet in the gap between Mars and Jupiter where the asteroids are now. They called themselves the Celestial Police and began to look for it. But they were beaten by an Italian called Guiseppe Piazzi. He discovered Ceres, the very first asteroid to be found. He could at least look a bit happier about it.

OCTOBER 2020

NASA's Osiris-Rex probe landed on an asteroid called Bennu and scooped up some rock samples. Osiris -Rex is now on its way back to Earth to deliver its cargo. Actual stuff from an asteroid brought back to Earth from over 300,000,000 miles (483,000,000 kms) away! That's some delivery service.

SO WHY GO THERE ?

Because they are full to the brim with the sort of minerals that we need back here on Earth. One day we'll have a whole mining industry on the asteroids, and when that happens, people will truly be living and working in outer space.

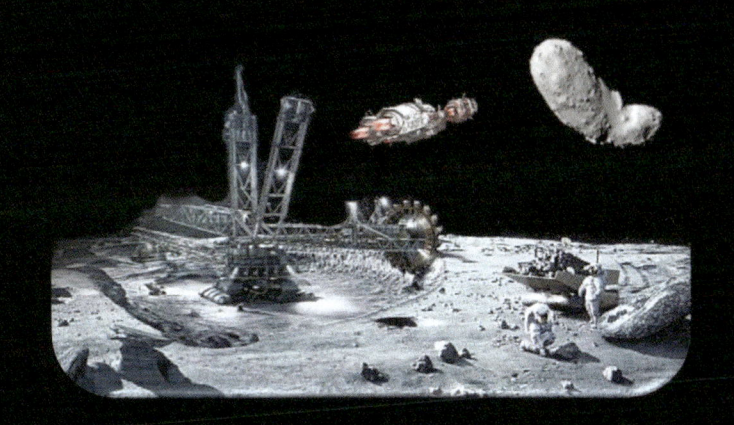

MESSENGER OF THE GODS

People in ancient times never saw asteroids in the sky unless they were crashing into the Earth (usually small ones called meteors). They thought that these were messengers from the gods and got really spooked when one appeared.

JUPITER

IT'S HUGE, REALLY HUGE

If you took all of the other planets in the solar system and squashed them together, Jupiter would still be 2½ times bigger than the planet that you end up with. It's the king of the solar system.

That big red spot in the photo? That's called - surprise, surprise – The Great Red Spot. It's a ginormous storm that's been raging for hundreds of years.

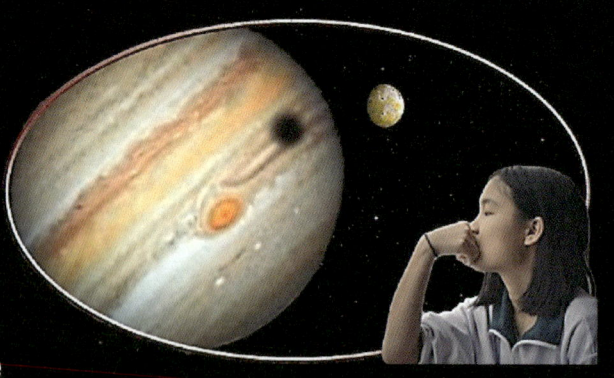

Jupiter is so big that it's gravity pulls in asteroids, comets and space junk which are destroyed when they hit the planet. In a real way, Jupiter helps protect all the other planets from getting bombarded. It's like having a giant vacuum cleaner in the solar system. Without this huge planet, our neighborhood would be a very different place. For one thing, all of the planets would go around the sun in different orbits from what they do now.

THE MOST BEAUTIFUL CLOUDS IN THE SOLAR SYSTEM

Jupiter's clouds are nearly 2,000 miles (3,200 kms) deep. That's right, 2,000 miles. As you can see, these clouds are all sorts of wonderful colors and shapes, just wonderful.

IT'S A GAS GIANT

A what? A gas giant is pretty much what it says – Jupiter is mostly just a huge ball of different gases. The other gas giants are Saturn, Uranus and Neptune. So if you wanted to find somewhere to stand on Jupiter, you'd fall through thousands of miles of gas before you got to anywhere solid. Not great for tourism.

JUPITER HAS RINGS AS WELL

They're hard to see, there's not many of them and they're no way as impressive as Saturn's, but they *are* there. So, Jupiter and Saturn both have ring systems and not a lot of people know that.

THE GALILEAN MOONS

Way back in 1610, a man in Italy called Galileo turned his homemade telescope (yep, homemade) toward Jupiter and discovered that there were four moons around it. These are the Galilean moons and they are easy to see even with a small telescope. Let's take a closer look at three of them.

CALLISTO

has the most craters of any other body in the solar system. Look at it, it's had a real bashing over millions of years.

IO

has over 400 active volcanoes, that's why it's red and yellow and is sometimes called the pizza moon. Io was the name of an ancient Greek woman who was turned into a cow by the god Zeus. Which was a pretty weird thing to do.

EUROPA

is about the same size as Earth's moon. The exciting thing is that it has oceans hidden under 10-15 miles (16 – kms) of ice. That water finds its way to the surface and bursts out as geysers.

Distance from the sun - 483 million miles (778 million kms)

Length of Day - 10 hours

Length of year – 12 Earth years

Diameter - 89,000 miles (143,000 kms)

Number of moons - 79

KING OF THE GODS

Jupiter is the name of the Roman king of the gods, that's why the biggest planet in our solar system is named after him. That's a lightning bolt in his hand - he really was the guy in charge.

A DUMB BUT FUN FACT

If Jupiter was a candy jar, you could fit 1,321 Earths inside it. Now, didn't you really need to know that?

A COMET HITS – LIVE!

In 1992, comet Shoemaker-Levy Nine broke up and smashed into Jupiter. Those dark swirls are where the pieces of comet hit the planet. Scientists were over the moon to get live photos of the bombardment.

SATURN

WOW! JUST WOW!

Is there anything more beautiful in the entire solar system than Saturn in all its glory? Answer: no.

WHAT ARE THE RINGS MADE OF?

Mostly bits of ice with some dust scattered in amongst them. There are huge boulder - sized chunks all through the rings, but most of the ice is actually quite small.

DINOSAURS AND RINGS?

Saturn's rings were made a long time after the planet was formed. They are only (only!) 100 million years old, so they were made when dinosaurs were still roaming around on Earth.

RAINING DIAMONDS?

Scientists think that it might be raining diamonds on Saturn. This is because the lightning in the atmosphere acts with a gas called methane and the soot that then falls turns hard and creates diamonds. They fall into the hot center of the planet and melt and it starts all over again.

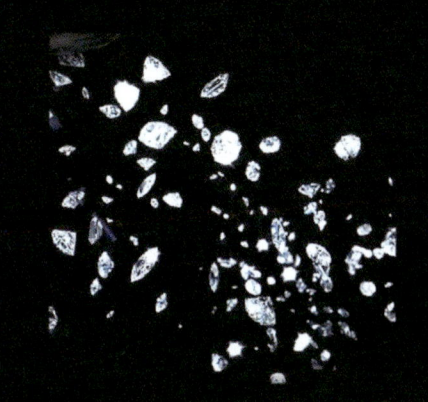

IT COULD FLOAT ON WATER

Saturn - like Jupiter, Uranus and Neptune - is mostly made up of gas. A lot of that is helium, the stuff you put in balloons. So the planet is very light and would actually float on water.

THE FLATTEST PLANET

Saturn spins really, really fast. Its day is only 10 hours long. Which is not long at all for such a big planet. Because of that, and the fact that it's made up mostly of gas, the planet is not really round. It's a little bit flat, sort of like a ball that's just a bit squashed

SATURN HAS MORE MOONS THAN JUPITER

So far, 53 have been named and another 20 are waiting to be given an official title. Let's take a look at some of them.

PITA BREAD MOON

This is the closest moon to Saturn that has a name. It's only about twenty miles (32 kms) across and is called Pan. It also looks a bit like ravioli pasta, doesn't it?

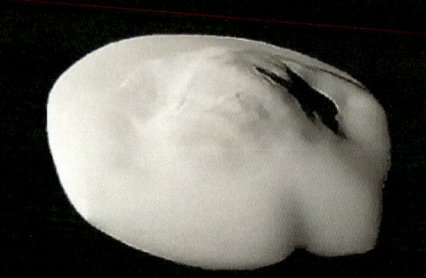

I NEED A NAME MOON

This moon was discovered in 1980 from the photographs taken by the Voyager space probe. It used to be called S/1980 S 28, but scientists came to their senses and now call it Atlas.

DEATH STAR MOON

This is Mimas. It was discovered by William Herschel in 1789, long before Star Wars became a thing. Spooky, right?

I CAN FLY MOON

Titan's atmosphere is so thick and its gravity so small, that if you strapped wings to your arms, you really could fly. True.

Distance from the sun - *889 million miles(14,300 million kms)*

Length of Day - *11 Hours*

Length of year - *29 Earth years*

Diameter – *74,900 miles (120,500 kms)*

Number of moons - *82*

WHO WAS SATURN ?

He was the god of farming, wealth and general good times. The Romans thought of him as a father of the gods and called his reign a golden age, a time of plenty and peace. But to us, we mostly only know that Saturday is named after him. Saturn's day.

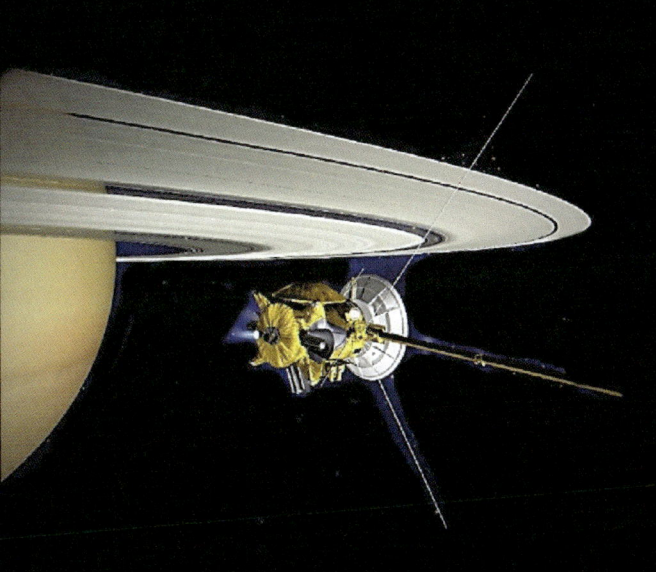

WELL DONE CASSINI !

From 2004 to 2017, the Cassini space probe sent 13 years' worth of images and data back to Earth. That's why we have lots of really cool photos of the rings and the planet.

DID SATURN MAKE CHRISTMAS?

In Ancient Rome, they had a festival to the god Saturn. This was a time for making merry, giving gifts and being with family. When did they do all of that? In December. So Saturnalia (which is what it was called) was a very early sort of Christmas.

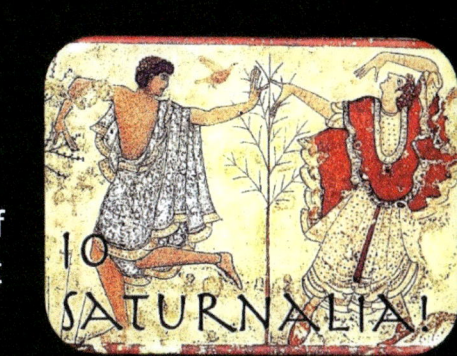

URANUS

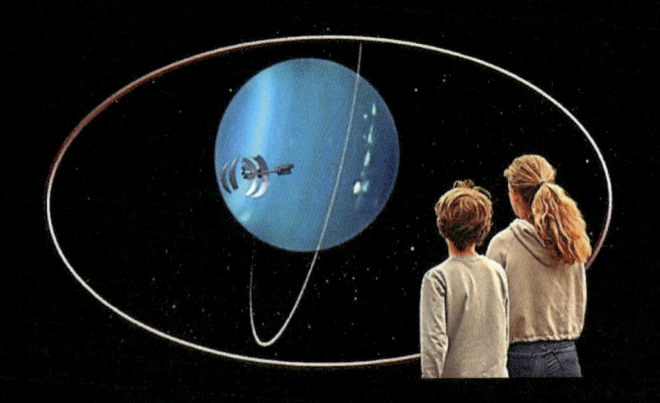

THE SIDEWAYS PLANET

Just when you thought the solar system couldn't get any weirder, along comes Uranus. It's sometimes called the sideways planet because that's how it spins - on its side. Uranus may have ended up that way because long ago another planet crashed into it.

ANOTHER ONE WITH RINGS

Along with Jupiter and Saturn, Uranus also has rings, 13 in fact. And because there's no dust in amongst them - not like on Jupiter and Saturn - they're almost impossible to see unless you take a special photo like this one.

OH, NO, NOT AGAIN !

Cupid and Belinda are two moons of Uranus that scientists say will crash into each other one day. But they'll be waiting a long time; the collision isn't going happen for about a million years.

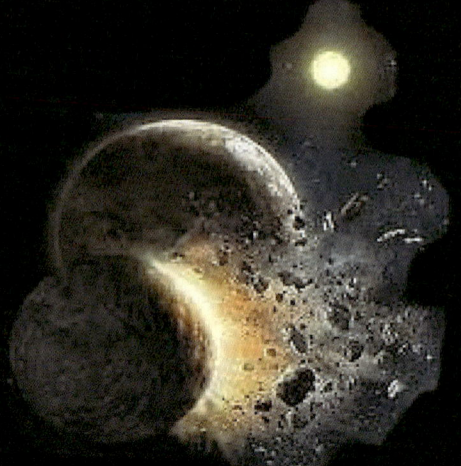

THE COLDEST OF THE COLD

It's not the farthest planet from the sun, but it is the coldest. Like all the gas giant planets (Jupiter, Saturn, Uranus and Neptune) its temperature is taken from the top of its clouds, which is counted as the surface. Uranus comes in as the coldest at *minus 375ºF.*

THE WINTER IS 21 YEARS LONG

Because the planet is tilted on its side, it has pretty weird seasons. It's like there are just two – summer/winter and spring/fall. During summer/winter, one side of the planet never sees the sun and the other side rarely sees the dark. During spring/fall, things are almost normal. Then there's a sort of day that lasts for around 17 Earth hours.

HUGE STORM

This picture of the planet's north pole shows a huge, huge storm. The big white patch? That's it. They don't get much bigger. Scientists think that the planet might be stormy because of the way it's tilted on its side, but they're not sure. It's another solar system mystery that we may never know the answer to.

UGH! ROTTEN EGGS

The clouds of Uranus are made of hydrogen sulphide, which is exactly the same gas that makes rotten eggs smell. So as well as being weird, Uranus is seriously smelly.

MOONS FROM THE PLAYS

Uranus has 27 known moons and most of them are named after people in plays by William Shakespeare. So we have moons with names like; Puck, Juliet, Miranda, Mab, Rosalind and Margaret. Which is a bit odd and not very moon-like, really.

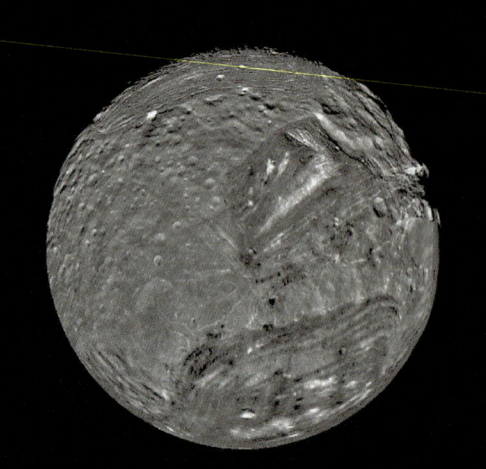

A JIGSAW GONE WRONG

When Voyager 2 sent back pictures of Miranda in 1986, scientists were shocked at what they saw. It was as if the moon had been broken apart and then put back together all wrong. There are cracks and ridges, canyons and cliffs and smooth plains, all jammed together like a really bad jigsaw puzzle.

WHY SO BLUE?

Uranus is a gas giant and its atmosphere is mostly made up of methane, helium and hydrogen. The methane gas reflects all the blue light waves that hit the planet, giving it this lovely color.

Distance from the sun - *1,800 million miles(2,900 million kms)*

Length of Day - *18 hours*

Length of year - *84 Earth years*

Diameter – *32,700 miles (52,000 kms)*

Number of moons – *27*

THE DADDY OF THE THEM ALL

In ancient Greek, Uranus meant 'sky', or 'father sky' and he was the father of all the gods. Uranus is the only planet to be named after a Greek god instead of a Roman one.

VOYAGER 2, OUR FARTHEST SPACECRAFT

In 1986, the Voyager 2 spacecraft swooped past and returned the first close-up images of the planet, its moons, and rings. Voyager 2 is now more than a billion miles (1.6 billion kms) away.

THE FIRST BY TELESCOPE

Uranus was the first planet to be discovered using a telescope. William Herschel was the astronomer who found it. This is a picture of his biggest telescope. When Uranus was discovered, it was the first new planet to be found in 2,000 years

NEPTUNE

THE END OF THE LINE

So now we are visiting the last planet in our solar system, Neptune. It may be the farthest from the sun, but it's no less wonderful than any of the others.

YET MORE RINGS

Neptune is yet another planet that has rings. There are five of them and they are more like the ones around Uranus than the ones around Saturn. And, yep, astronomers think that this is the result of yet another collision in the solar system. This time, probably between two moons that were circling the planet.

BIGGER, BUT YOU'D WEIGH THE SAME

Neptune may be 4 times bigger than Earth, but its gravity is about the same. In fact, you'd weigh pretty much what you do here on Earth. It's that gas giant thing again: lots of stuff spread thin over a big area, so it doesn't weigh as much as it looks like it should.

A GREAT DARK SPOT

Remember the Great Red Spot on Jupiter? Well, this was the same sort of deal. It was a huge storm about the size of Earth. It faded away and in 5 years was gone. But because of the way Neptune is made up, there will be others like it for millions of years to come.

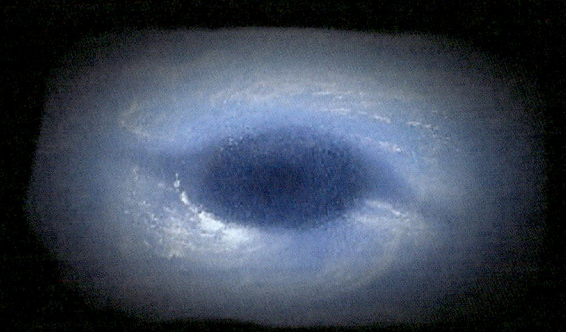

Neptune has storms like the Dark Spot because it's the *windiest* planet in the solar system. In fact, the Great Dark Spot raced along with winds of over 1,500 miles (2,400 kms) per hour.

'I DID! ' 'NO, I DID!

In the 1800s, two men argued about who found Neptune first, and the credit was given to the French astronomer. He wanted it named after himself, but no-one else did. Then he suggested calling it Neptune instead. Which was just as well, because his name was Le Verrier.

THE BACKWARD MOON

Triton is the biggest of Neptune's moons. It's different from all the others in the entire solar system because it orbits opposite to the way that Neptune spins.

Distance from the sun - *2,800 million miles(4,500 million kms)*

Length of Day - *16 hours*

Length of year - *165 Earth years*

Diameter - *30,800 miles (50,000 kms)*

Number of moons - *14*

GOD OF THE OCEANS

To the Greeks he was Poseidon and to the Romans he was Neptune. He was the god of the seas and fresh water – streams and rivers and things like that. He was also the god of horses and horse racing.

THE SEAHORSE IN SPACE

In 2013 astronomers found a tiny moon close to Neptune. It's only 21 miles (34 kms) wide and was incredibly hard to spot. One of the team who found it called the moon Hippocamp, which means sea horse.

SO FAR, FAR, AWAY

The sunlight we receive on Earth is about 900 times brighter than that which falls on Neptune. The farthest planet from the sun is so far away that we're not going to land on it anytime soon. But in the far future? Definitely.

So that's the end of our tour of the planets in our solar system. As you have seen, each of our seven neighbors in space is different, special and in its own way, wonderful. To enjoy the magnificence of where we live in this universe, all you need is a dark night, to look up and to never stop dreaming.

Printed in Great Britain
by Amazon